# Born Inside a Nest So Small

## Conversation with a Hummingbird

~ Iza Trapani ~

Charlesbridge

Hummingbird, you nectar lover,
how you dash and dart and hover—
forward, backward, upside down.
I love to watch you zip around.

Hummingbirds are the
smallest birds in the world.
There are many species.

This is a ruby-throated hummingbird. It is smaller than an ice-cream stick!

See this book I'm looking through?
Now here's a fact—I learned that you
were born inside a nest so small
it wouldn't fit a Ping-Pong ball.

Male ruby-throated hummingbirds have red throats, but females do not.

In the nest your mama made,
a teeny, tiny egg she laid.

A ruby-throated hummingbird's eggs are the size of jelly beans.

From that egg you hatched one day,
then three weeks later, flew away.

Whizzing by me, you're a blur
with bitty wings that *whir whir whir.*
Every second—*flap flap flap*—
fifty beats or more you clap!

Hummingbirds' beating wings make the humming sound that gives them their name.

In my garden, full in bloom,
to the hollyhocks you zoom.
Sipping nectar all day through,
munching on a bug or two.

Hummingbirds eat gnats,
fruit flies, and mosquitoes,
in addition to nectar from flowers.

Hummingbird feeders
are filled with a mixture
of sugar and water.

But to me it's even sweeter
when I spot you at my feeder.
There you sip so happily,
while keeping careful watch on me.

Sneaky little needle-beak,
are you playing hide-and-seek?
You're a perky little one,
and seeing you is so much fun!

Hummingbirds can't walk,
but they can grip with
their tiny feet.

Now that fall is drawing near,
you must travel south of here.
A thousand miles or more you'll fly.
It's time for us to say goodbye.

As summer ends, hummingbirds
migrate to warmer places like
southern Mexico or Central America.

Bye, my friend, and off you go,
flying steady, flying low.

Spotting snacks along your trip—
here a nibble, there a sip.

Hummingbirds fly alone over great
distances, around 1,500 miles
(2,400 kilometers) or more!

Over water, whizzing, whooshing,
hours later, you'll keep pushing.
You're a small but sturdy flier,
though I'm sure you're bound to tire.

Hummingbirds can fly nonstop
for nearly twenty-four hours when
crossing over the Gulf of Mexico.

Still you'll brave the rain and thunder,
lightning, headwinds—you're a wonder!
Speeding through the storms and showers
till you reach the sun and flowers.

When you feel the warming sun,
then at last your trip is done.
Finally you'll have a rest.
Enjoy a nectar-sipping fest!

Hummingbirds sip nectar
through long tongues
inside their hollow beaks.

With your feathers all aglitter,
you will flit and fly and flitter.

Hummingbirds migrate twice
a year. They often take the same
route and return to the same place.

In your winter home you'll stay,
until you head back north one day.

How I wish that you could send
a postcard when you're there, my friend.
Travel safely, little hummer.
Let's meet up again next summer.

## Author Note

In early fall one year, I was looking out my kitchen window, thinking it was time to put away the hummingbird feeder. I hadn't seen the tiny birds for a few days and thought they must've been on their way south to warmer lands.

At that moment, a male ruby-throated hummingbird appeared at the feeder. He took a quick sip, then flew toward the open window where I stood watching. Much to my surprise, the hummingbird flew right up to the screen. There he hovered and cocked his head left and right, staring at me.

After a few seconds, he flew backward, took another sip from the feeder, looked at me again, and flew away. I was amazed. Had he come to thank me for the nectar? Was he saying goodbye? This spunky little bird inspired me to write a poem that later would be made into this book.

## Recommended Books

Burleigh, Robert. *Tiny Bird: A Hummingbird's Amazing Journey.*
New York: Henry Holt, 2020.

Shewey, John. *The Hummingbird Handbook: Everything You
Need to Know About These Fascinating Birds.* Portland, OR:
Timber Press, 2021.

Sill, Cathryn. *About Hummingbirds: A Guide for Children.*
Atlanta, GA: Peachtree, 2011.

## Recommended Websites

www.allaboutbirds.org
www.audubon.org
www.hummingbirdcentral.com
www.journeynorth.org

*For Harry, with love*

Charlesbridge · 9 Galen Street, Watertown, MA 02472
www.charlesbridge.com

**Library of Congress Cataloging-in-Publication Data**
Names: Trapani, Iza, author. | Trapani, Iza, illustrator.
Title: Born inside a nest so small: conversation with a hummingbird / Iza Trapani.
Description: Watertown, MA: Charlesbridge, [2025] | Audience: Ages 4–7 | Audience: Grades K–1 | Summary: "A young child observes and admires a friendly ruby-throated hummingbird in this poetic love letter to the smallest birds in the world."—Provided by publisher.
Identifiers: LCCN 2024042611 (print) | LCCN 2024042612 (ebook) | ISBN 9781623545840 (hardcover) | ISBN 9781632894557 (ebook)
Subjects: LCSH: Ruby-throated hummingbird—Juvenile literature. | Hummingbirds—Juvenile literature.
Classification: LCC QL696.A558 T73 2025 (print) | LCC QL696.A558 (ebook) | DDC 598.7/64—dc23/eng/20241118
LC record available at https://lccn.loc.gov/2024042611
LC ebook record available at https://lccn.loc.gov/2024042612

Printed in China · OPIC
(hc) 10 9 8 7 6 5 4 3 2 1

Illustrations done in watercolor and acrylic gouache paints, colored pencil, and pastels on Arches watercolor paper
Text type set in Blauth
Designed by Diane M. Earley
Production supervised by Jennifer Most Delaney